Detectives: Animal Smarts

Search for the Facts...

Using Tools

Anne O'Daly

BROWN BEAR BOOKS

Published by Brown Bear Books Ltd
4877 N. Circulo Bujia
Tucson, AZ 85718
USA

and

G14, Regent Studios
1 Thane Villas
London N7 7PH
UK

ISBN 978-1-83572-028-8 (ALB)
ISBN 978-1-83572-034-9 (paperback)
ISBN 978-1-83572-040-0 (ebook)

Library of Congress Cataloging-in-Publication Data available on request

Design Manager: Keith Davis
Children's Publisher: Anne O'Daly
Picture Manager: Sophie Mortimer

Picture Credits
Cover: Shutterstock: Badrudin15. Interior: Alamy: Steve Bloom Images 8–9; Dreamstime: Agami Photo Agency 18–19; iStock: CR Bellette 5c, curious tiger 9, Akarat Duangkhong 5b, jeryltan 5tl, Mike Lane45 4–5c, Ian Stych 12¬–13, Vladimir Vladimirov 1, 9; Shutterstock: Joe Avison 6, Bernard Beaussier 20–21, Lisa Crawford 16–17, ferryelegant 6–7, Russ Jenkings 18, Mike Lane45 23, Agata Marianna 14, Mshake 4t, ROBERT67 4–5b, 10–11, Menno Schaefer 22tr, SergeUWPhoto 4b, 14–15, Andrey Shcherukhin 22tl, Luke Shelley 22bl, Dai mar Tamarack 20, Alec Tursler 12, Sergey Uryadnikov 22br, Ashley Wieren 16, Hubert Yann 5tr.
t=top, b=bottom, l=left, r=right, c=center
All artwork and other photography Brown Bear Books.

Brown Bear Books has made every attempt to contact the copyright holder.
If you have any information about omissions, please contact: licensing@brownbearbooks.co.uk

Manufactured in the United States of America
CPSIA compliance information: Batch#AG/5663

Websites
The website addresses in this book were valid at the time of going to press. However, it is possible that contents or addresses may change following publication of this book. No responsibility for any such changes can be accepted by the author or the publisher. Readers should be supervised when they access the Internet.

Contents

Using Tools

**We use tools to help us do jobs.
Many other animals use tools, too.
They use them to get food or for shelter.
Meet some amazing animal tool-users!**

Orangutan

Capuchin Monkey

Coconut Octopus

Sea Otter

Using Tools

Chimp

Dolphin

Bowerbird

Bowerbirds decorate their nests with shiny objects.

Crow

Orangutan

Orangutans are clever apes. They use long sticks to get fruit. They use leaves in clever ways, too.

FACT FILE

Name: orangutan

Where it lives: rainforests in Indonesia and Malaysia

What it eats: fruit and leaves

Type of animal: mammal

Size: 3.5 to 4.5 feet (1 to 1.4 m)

Body: long arms, red brown fur

Orangutans sleep high up in trees. They make nests from leaves and branches.

MINI FACTS

Orangutans make a glove out of leaves. They use it to pick prickly fruit.

Chimp

Chimps have different tools for different jobs. They use stones to open nuts. They fish for insects with sticks.

FACT FILE

Name: chimpanzee

Where it lives: rainforests in Africa

What it eats: mainly fruit, leaves, nuts, and insects

Type of animal: mammal

Size: up to 5 feet (1.5 m)

Body: black fur and long arms

MINI FACTS

A chimp called Washoe learned sign language. She knew 350 words!

A chimp uses a stick to get termites

Babies learn how to use tools from their moms.

Sea Otter

Sea otters feed on shellfish. Shellfish have hard shells. The otters use stones to break the shells. They eat the soft meat inside.

FACT FILE

Name: sea otter

Where it lives: in oceans near the coast

What it eats: clams, sea urchins, and fish

Type of animal: mammal

Size: 4 feet (1.2 m)

Body: webbed feet for swimming; thick fur to keep warm

Young otters learn to use a stone from their mom

MINI FACTS

Some otters have a favorite stone. They use it til it wears out.

Baby sea otters are born in the water. But they can't swim at first. The mom carries her baby on her chest.

Bowerbird

Bowerbirds build nests called bowers. They decorate the nests with flowers. Males make nests to find a mate.

FACT FILE

Name: satin bowerbird

Where it lives: Australia

What it eats: fruit, seeds, and insects

Type of animal: bird

Size: 11 to 14 inches (27 to 33 cm)

Body: males have shiny black feathers, females are brown

Some birds collect human rubbish

The birds collect berries and feathers

Golden bowerbirds line their nest with flowers.

The female chooses the best nest

MINI FACTS

Some bowerbirds chew berries. They make a paste. They use it to paint their nest.

Coconut Octopus

This clever animal lives in oceans. The octopus picks up a clam shell. It uses the shell for a shelter. The shell keeps the animal safe.

FACT FILE

Name: coconut octopus

Where it lives: Pacific Ocean

What it eats: small fish and shrimp

Type of animal: mollusc

Size: body is 3 inches (8 cm) long; arms are 6 inches (15 cm) long

Body: dark body with stripes; eight legs

People throw coconut shells into the ocean. The octopuses use the shells for shelter.

MINI FACTS

The octopus cleans its shell with a jet of water.

The octopus hides inside the shell

It carries the shell when it moves

Capuchin Monkey

Capuchin monkeys dig for spiders. They use stones to help them dig. The monkeys use sticks to fish for insects.

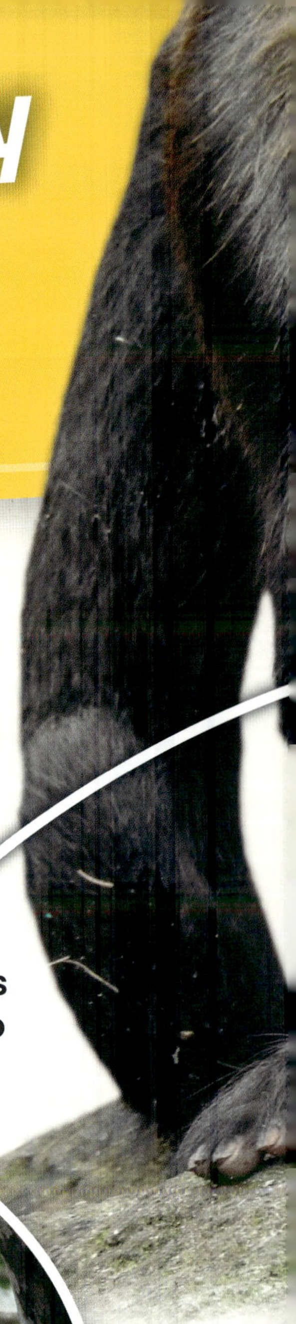

The monkeys use stones to open nuts

FACT FILE

Name: capuchin monkey

Where it lives: forests in South and Central America

What it eats: fruit, insects, leaves, small birds, and nuts

Type of animal: mammal

Size: 12 to 22 inches (30 to 56 cm)

Body: white fur around the face, long tail

A capuchin mom carries her baby on her back.

MINI FACTS

Capuchin monkeys throw rocks at predators. This scares the predators away.

Crow

Crows make tools from twigs and leaves. They bend a twig to make a hook. They poke it into a tree to find insects.

FACT FILE

Name: crow

Where it lives: around the world, in woodlands and cities

What it eats: plants and small animals

Type of animal: bird

Size: 8.5 to 26 inches (21.5 to 65 cm)

Size: shiny black feathers

The twig gets to places that are hard to reach

Crows in cities drop nuts onto a road. Cars drive over the nuts. The cars crack the nuts open.

MINI FACTS

Crows have favorite twigs. The birds save these twigs in a safe place.

Dolphin

Some dolphins put a sponge over their face. Then they look for food on the seafloor. The sponge saves the dolphin from scratches.

FACT FILE

Name: bottlenose dolphin

Where it lives: warm seas around the world

What it eats: fish, squid, and shrimp

Type of animal: mammal

Size: up to 13 feet (4 m) long

Body: smooth skin, flippers for swimming

Dolphins breathe through a hole on their head. The hole is called a blowhole.

MINI FACTS

Dolphins sleep with one eye open. Half their brain is awake all the time.

Quiz

Test your skills!
Can you answer these questions?
The answers are on page 24.

1 Where do orangutans sleep?

2 Where do sea otters keep their favorite stone?

3 Why do bowerbirds make nests?

4 How do dolphins breathe?

Useful Words

ape A type of animal that is closely related to humans. Chimps and orangutans are apes.

blowhole A hole that whales and dolphins have on the top of their head.

bower The nest a bowerbird makes to attract a female.

burrow A hole that an animal digs for a home.

mammal A type of animal that has fur and makes milk to feed their babies.

nest A structure made of leaves and twigs.

paste A soft, wet mixture.

predator An animal that hunts and eats other animals.

sponge An animal that lives in the oceans.

Find Out More

Books

Chimpanzees (Amazing Animals), Kate Riggs (The Creative Company, 2022)

Dolphins: Learned Behaviors, Laurie Allen Klein (Arbordale Publishing, 2024)

Smart Animals, Michael Holland (Little Gestalten, 2023)

Websites

www.coolkidfacts.com/sea-otters-facts/

www.smorescience.com/top-10-smartest-animals/

www.natgeokids.com/uk/discover/animals/general-animals/10-chimpanzee-facts/

Index

Quiz Answers: 1. Orangutans sleep in nests high up in trees; **2.** They keep the stone under their arm; **3.** Male bowerbirds make nests to find a mate; **4.** Dolphins breathe through a hole on their head.